DINO'S BUBBLE TROUBLE

ALEX KALEDORA

"Dino's Bubble Trouble"
Copyright © 2025 Alex Kaledora

All rights reserved. No part of this book may be reproduced, stored in a retrieval system, or transmitted in any form or by any means, electronic, mechanical, photocopying, recording, or otherwise, without the prior written permission of the publisher, except in the case of brief quotations embodied in critical articles and reviews.

One sunny morning, Dino the dinosaur found a shiny pink gumball on the ground. "Ooooh, candy!" he cheered, popping it into his mouth and chomping with a CRUNCH-CRUNCH-CHOMP.

Suddenly — POP! — a bubble burst from his nose! Dino blinked and giggled. "Well, that was snorty-fun!"

Dino tried again, blowing his biggest bubble yet — it grew and grew like a pink balloon. "Look at me!" he roared, cheeks puffing, as the bubble got as big as a watermelon.

But then… whoosh! The bubble lifted Dino's feet off the ground! "Uh-oh!" Dino squeaked. "I'm flying!"

Up, up, up Dino floated, wobbling like a giant pink balloon-animal. Birds flapped past, chirping, "What's THAT big thing in the sky?!"

"Hi, birdies!" Dino called, waving a tiny arm. "I'm Dino! I'm... uh... sort of stuck."

Below, Steggy the stegosaurus looked up and gasped. "Dino! You're a flying lollipop!" Steggy shouted.

Dino giggled, "I'm not a lollipop, I'm a DINOSAUR-POP!" But secretly, he wondered how to get down.

The wind blew harder, carrying Dino over the playground. The dino kids looked up, jaws wide open. "Dino's in the clouds!" they cheered, clapping their tiny claws.

"Help me down, please!" Dino called, but the kids just waved and shouted, "Do a loop-de-loop!"

As Dino floated over the lake, he spotted his reflection in the water. "Whoa, I look like a giant jellybean with legs!" Dino snorted, nearly popping the bubble from laughing.

A family of ducks quacked up at him, "Quack-quack, nice hat!" Dino chuckled, "Thanks, quackers!"

Suddenly — SNAP! A branch poked Dino's bubble. "Uh-oh…" Dino gulped. The bubble wiggled. The bubble jiggled. The bubble… held on.

"Phew!" Dino sighed. "Close call!" But just then, a butterfly landed on the bubble — POP!

Dino tumbled down, arms flapping like a baby chicken. "Aaaah!" he yelped as he crashed right into a haystack with a soft PLOOF!

Straw flew everywhere, covering Dino's head like a fluffy yellow wig. "Ta-da!" Dino grinned. "Fashion forward!"

His friends rushed over. "Dino! That was the coolest ride EVER!" Steggy said, eyes wide. "Are you okay?"

Dino pulled hay from his teeth and smiled. "I'm good! But maybe no more bubble rides today."

Just then, Dino felt something wiggly in his mouth. Another bubble! "Oh no, here we go again!" Dino giggled nervously.

The bubble grew and grew until it lifted his toes off the ground again. "Somebody grab my tail!" Dino yelped.

A friend leapt and chomped Dino's tail. Followed by the others.

"Hold on, team!" Dino shouted. But the bubble pulled... and pulled... until WHOOSH! Everyone shot up like a dino-kite string.

Up they floated, dinosaurs dangling like popcorn on a string. "Wheeeeee!"

"This is the silliest parade EVER!" giggled Dino, swinging like a Christmas ornament.

Finally, Dino scrunched up his face, gave one big CHOMP — and popped the bubble himself.

The whole dino-chain tumbled into a giant leaf pile with a loud, happy KER-FLOOF!

Dino poked his head out of the leaves and grinned. "Okay, maybe bubbles are fun... but next time, let's stick to bubble popping!"

All the dinos laughed and tossed leaves in the air, shouting, "BEST. DAY. EVER!"

The End.

www.ingramcontent.com/pod-product-compliance
Lightning Source LLC
LaVergne TN
LVHW071655060526
838200LV00030B/472